Higher Mathematics

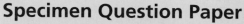

Specimen Question Paper
Units 1, 2 and 3 Paper 1 (Non-calculator)
Units 1, 2 and 3 Paper 2

2000 Exam
Units 1, 2 and 3 Paper 1 (Non-calculator)
Units 1, 2 and 3 Paper 2

2001 Exam
Units 1, 2 and 3 Paper 1 (Non-calculator)
Units 1, 2 and 3 Paper 2

2002 Exam
Units 1, 2 and 3 Paper 1 (Non-calculator)
Units 1, 2 and 3 Paper 2

2002 Winter Diet Exam
Units 1, 2 and 3 Paper 1 (Non-calculator)
Units 1, 2 and 3 Paper 2

2003 Exam
Units 1, 2 and 3 Paper 1 (Non-calculator)
Units 1, 2 and 3 Paper 2

© Scottish Qualifications Authority

All rights reserved. Copying prohibited. No part of this publication may be reproduced, stored in a retrieval system, or transmitted in any form or by any means, electronic, mechanical, photocopying, recording or otherwise.

First exam published in 1999.
Published by
Leckie & Leckie, 8 Whitehill Terrace, St. Andrews, Scotland KY16 8RN
tel: 01334 475656 fax: 01334 477392
enquiries@leckieandleckie.co.uk www.leckieandleckie.co.uk

Leckie & Leckie Project Team: Peter Dennis; John MacPherson; Bruce Ryan; Andrea Smith

ISBN 1-84372-126-0

A CIP Catalogue record for this book is available from the British Library.

Printed in Scotland by Scotprint.

Leckie & Leckie is a division of Granada Learning Limited, part of Granada plc.

Scotland's leading educational publishers

Introduction

Dear Student,

This past paper book provides you with the perfect opportunity to put into practice everything you should know in order to excel in your exams. The compilation of papers will expose you to an extensive range of questions and will provide you with a clear idea of what to expect in your own exam this summer.

The past papers represent an integral part of your revision: the questions test not only your subject knowledge and understanding but also the examinable skills that you should have acquired and developed throughout your course. The answer booklet at the back of the book will allow you to monitor your ability, see exactly what an examiner looks for to award full marks and will also enable you to identify areas for more concentrated revision. Make use too of the tips for revision and sitting your exam to ensure you perform to the best of your ability on the day.

Practice makes perfect. This book should prove an invaluable revision aid and will help you prepare to succeed.

Good luck!

Acknowledgements

Every effort has been made to trace the copyright holders and to obtain their permission for the use of copyright material. Leckie & Leckie will gladly receive information enabling them to rectify any error or omission in subsequent editions.

HIGHER SPECIMEN QUESTION PAPER

X100/301

NATIONAL
QUALIFICATIONS

Time: 1 hour 10 minutes

**MATHEMATICS
HIGHER
Units 1, 2 and 3
Paper 1
(Non-calculator)**

SPECIMEN QUESTION PAPER

Read Carefully

1 **Calculators may NOT be used in this paper.**

2 Full credit will be given only where the solution contains appropriate working.

3 Answers obtained by readings from scale drawings will not receive any credit.

FORMULAE LIST

Circle:

The equation $x^2 + y^2 + 2gx + 2fy + c = 0$ represents a circle centre $(-g, -f)$ and radius $\sqrt{g^2 + f^2 - c}$.

The equation $(x - a)^2 + (y - b)^2 = r^2$ represents a circle centre (a, b) and radius r.

Scalar Product: $\quad \mathbf{a}.\mathbf{b} = |\mathbf{a}||\mathbf{b}| \cos \theta$, where θ is the angle between $\mathbf{a}$ and $\mathbf{b}$

or $\quad \mathbf{a}.\mathbf{b} = a_1 b_1 + a_2 b_2 + a_3 b_3$ where $\mathbf{a} = \begin{pmatrix} a_1 \\ a_2 \\ a_3 \end{pmatrix}$ and $\mathbf{b} = \begin{pmatrix} b_1 \\ b_2 \\ b_3 \end{pmatrix}$.

Trigonometric formulae:

$\sin(A \pm B) = \sin A \cos B \pm \cos A \sin B$
$\cos(A \pm B) = \cos A \cos B \mp \sin A \sin B$
$\sin 2A = 2 \sin A \cos A$
$\cos 2A = \cos^2 A - \sin^2 A$
$\quad\quad\quad = 2\cos^2 A - 1$
$\quad\quad\quad = 1 - 2\sin^2 A$

Table of standard derivatives:

$f(x)$	$f'(x)$
$\sin ax$	$a \cos ax$
$\cos ax$	$-a \sin ax$

Table of standard integrals:

$f(x)$	$\int f(x)\, dx$
$\sin ax$	$-\frac{1}{a} \cos ax + C$
$\cos ax$	$\frac{1}{a} \sin ax + C$

ALL questions should be attempted.

Marks

1. Show that $x = 2$ is a root of the equation $y = 2x^3 + x^2 - 13x + 6 = 0$ and, otherwise, find the other roots.

2. A and B are the points $(-3, -1)$ and $(5, 5)$.
   Find the equation of the perpendicular bisector of AB.

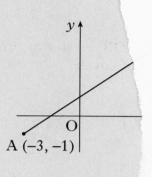

3. The point $P(-1, 7)$ lies on the curve with equation $y = 5x^2 + 2$. Find the equation of the tangent to the curve at P.

   3

4. The line AB makes an angle of $\frac{\pi}{3}$ radians with the y-axis, as shown in the diagram. Find the exact value of the gradient of AB.

   2

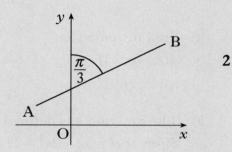

5. (a) The diagram shows a circle, centre P, with equation $x^2 + y^2 + 6x + 4y + 8 = 0$.
   Find the equation of the tangent at the point $A\,(-1, -1)$ on the circle.

   4

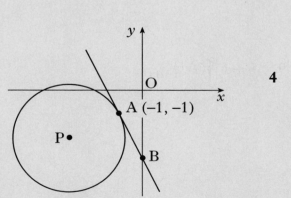

   (b) The tangent crosses the y-axis at B.
   Find the equation of the circle with AB as diameter.

   3

Marks

6. $f(x) = \sqrt{3}\sin x° - \cos x°$

   (a) Express $f(x)$ in the form $k\sin(x - a)°$ where $k > 0$ and $0 \leq a < 360$. 4

   (b) Hence solve the equation $f(x) = \sqrt{2}$ in the interval $0 \leq a < 360$. 3

7. Using triangle PQR, as shown, find the exact value of $\cos 2x$. 3

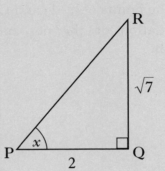

8. Functions f and g are defined on the set of real numbers by

   $$f(x) = x - 1$$
   $$g(x) = x^2.$$

   (a) Find formulae for
   (i) $f(g(x))$
   (ii) $g(f(x))$. 3

   (b) The function h is defined by $h(x) = f(g(x)) + g(f(x))$.
   Show that $h(x) = 2x^2 - 2x$ and sketch the graph of h. 3

   (c) Find the area enclosed between this graph and the x-axis. 4

9. Find $\int \dfrac{x^2 - 5}{x\sqrt{x}} dx$. 4

10. The diagram shows two vectors **a** and **b**, with |**a**| = 3 and |**b**| = 2√2. These vectors are inclined at an angle of 45° to each other.

(a) Evaluate (i) **a.a**

(ii) **b.b**

(iii) **a.b**

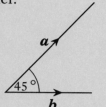

2

(b) Another vector **p** is defined by **p** = 2**a** + 3**b**.

Evaluate **p.p** and hence write down |**p**|.

4

[*END OF QUESTION PAPER*]

[BLANK PAGE]

X100/303

NATIONAL QUALIFICATIONS

Time: 1 hour 30 minutes

MATHEMATICS HIGHER
Units 1, 2 and 3
Paper 2

SPECIMEN QUESTION PAPER

Read Carefully

1 **Calculators may be used in this paper.**
2 Full credit will be given only where the solution contains appropriate working.
3 Answers obtained by readings from scale drawings will not receive any credit.

FORMULAE LIST

Circle:

The equation $x^2 + y^2 + 2gx + 2fy + c = 0$ represents a circle centre $(-g, -f)$ and radius $\sqrt{g^2 + f^2 - c}$.

The equation $(x - a)^2 + (y - b)^2 = r^2$ represents a circle centre (a, b) and radius r.

Scalar Product: $\quad \mathbf{a}.\mathbf{b} = |\mathbf{a}| |\mathbf{b}| \cos \theta$, where θ is the angle between $\mathbf{a}$ and $\mathbf{b}$

$\quad$ or $\quad \mathbf{a}.\mathbf{b} = a_1 b_1 + a_2 b_2 + a_3 b_3$ where $\mathbf{a} = \begin{pmatrix} a_1 \\ a_2 \\ a_3 \end{pmatrix}$ and $\mathbf{b} = \begin{pmatrix} b_1 \\ b_2 \\ b_3 \end{pmatrix}$.

Trigonometric formulae:

$\sin(A \pm B) = \sin A \cos B \pm \cos A \sin B$

$\cos(A \pm B) = \cos A \cos B \mp \sin A \sin B$

$\sin 2A = 2 \sin A \cos A$

$\cos 2A = \cos^2 A - \sin^2 A$

$\quad \quad \quad = 2\cos^2 A - 1$

$\quad \quad \quad = 1 - 2\sin^2 A$

Table of standard derivatives:

$f(x)$	$f'(x)$
$\sin ax$	$a \cos ax$
$\cos ax$	$-a \sin ax$

Table of standard integrals:

$f(x)$	$\int f(x)\,dx$
$\sin ax$	$-\frac{1}{a} \cos ax + C$
$\cos ax$	$\frac{1}{a} \sin ax + C$

ALL questions should be attempted.

1. The parabola shown in the diagram has equation $y = 4x - x^2$ and intersects the x-axis at the origin and P.

   The line PQ has equation $2y + x = 4$.

   Find the coordinates of P and Q.

   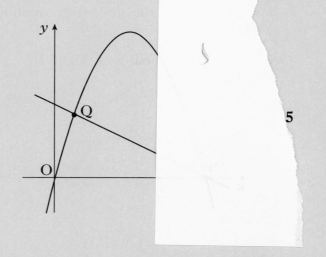

   5

2. OABCEFGH is a cuboid.

   With axes as shown, O is the origin and the coordinates of A, H, G and L are $(4, 0, 0)$, $(0, 0, 5)$, $(0, 6, 5)$ and $(3, 6, 5)$ respectively.

   K lies two thirds of the way along HG, (ie HK:KG = 2:1).

   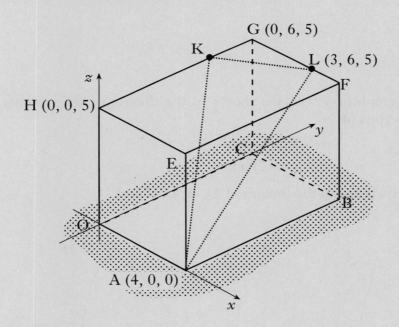

   (a) Determine the coordinates of K. 2

   (b) Write down the components of $\overrightarrow{AK}$ and $\overrightarrow{AL}$. 2

   (c) Calculate the size of angle KAL. 5

Marks

3. The diagram shows part of the graph of $y = 6\sin 3x$ and the line with equation $y = 4$.

   Find the x-coordinates of A and B.

   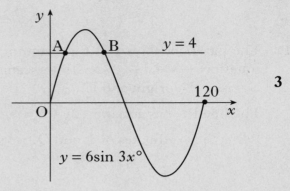

   3

4. Two sequences are defined by the recurrence relations

   $$u_{n+1} = 0\cdot 2u_n + p, \quad u_0 = 1 \quad \text{and}$$
   $$v_{n+1} = 0\cdot 6v_n + q, \quad v_0 = 1.$$

   (a) Explain why each of these sequences has a limit. **1**

   (b) If both sequences have the same limit, express p in terms of q. **3**

5. Part of the graph of $y = f(x)$ is shown in the diagram. On separate diagrams, sketch the graphs of

   (i) $y = f(x + 1)$

   (ii) $y = -2f(x)$.

   Indicate on each graph the images of O, A, B, C and D. **5**

   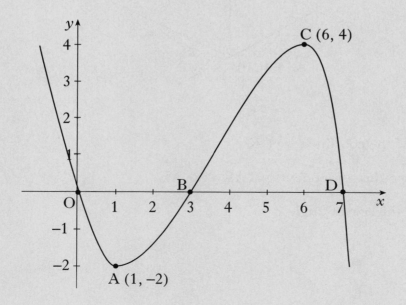

Marks

6. The graph of $y = g(x)$ passes through the point $(1, 2)$.

   If $\dfrac{dy}{dx} = x^3 + \dfrac{1}{x^2} - \dfrac{1}{4}$, express y in terms of x. 4

7. The intensity I_t of light is reduced as it passes through a filter [...] law $I_t = I_0 e^{-kt}$ where I_0 is the initial intensity and I_t is the inten[sity] through a filter of thickness t cm. k is a constant.

   (a) A filter of thickness 4 cm reduces the intensity from 120 [candle-power] to 90 candle-power. Find the value of k. 4

   (b) The light is passed through a filter of thickness 10 cm. Find the percentage reduction in its intensity. 3

8. The diagram shows part of the graph of $y = \log_b(x + a)$.

   Determine the values of a and b. 3

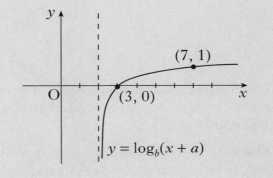

9. A curve has equation $y = 2x^3 + 3x^2 + 4x - 5$.

   Prove that this curve has no stationary points. 5

10. A zookeeper wants to fence off six individual animal pens.

Each pen is a rectangle measuring x metres by y metres, as shown in the diagram.

(a) (i) Express the total length of fencing in terms of x and y.

(ii) Given that the total length of fencing is 360 m, show that the total area, A m^2, of the six pens is given by $A(x) = 240x - \frac{16}{3}x^2$. **4**

(b) Find the values of x and y which give the maximum area and write down this maximum area. **6**

11. The diagram shows a sketch of the graph of $y = x^3 - 9x + 4$ and two parallel tangents drawn at P and Q.

The equations of the tangents at P and Q are $y = 3x + 20$ and $y = 3x - 12$ respectively.

Show that the shortest distance between the tangents is $\frac{16\sqrt{10}}{5}$. **5**

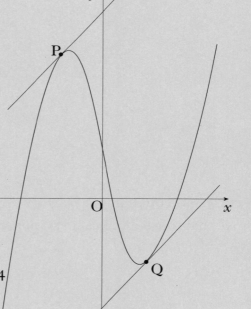

[END OF QUESTION PAPER]

X056/301

NATIONAL QUALIFICATIONS 2000

THURSDAY, 25 MAY 9.00 AM – 10.10 AM

MATHEMATICS
HIGHER
Paper 1
(Non-calculator)
Units 1, 2 and 3

Read Carefully

1 **Calculators may NOT be used in this paper.**

2 Full credit will be given only where the solution contains appropriate working.

3 Answers obtained by readings from scale drawings will not receive any credit.

FORMULAE LIST

Circle:

The equation $x^2 + y^2 + 2gx + 2fy + c = 0$ represents a circle centre $(-g, -f)$ and radius $\sqrt{g^2 + f^2 - c}$.

The equation $(x - a)^2 + (y - b)^2 = r^2$ represents a circle centre (a, b) and radius r.

Scalar Product: $\quad \mathbf{a}.\mathbf{b} = |\mathbf{a}||\mathbf{b}| \cos \theta$, where θ is the angle between $\mathbf{a}$ and $\mathbf{b}$

$\quad$ or $\quad \mathbf{a}.\mathbf{b} = a_1 b_1 + a_2 b_2 + a_3 b_3$ where $\mathbf{a} = \begin{pmatrix} a_1 \\ a_2 \\ a_3 \end{pmatrix}$ and $\mathbf{b} = \begin{pmatrix} b_1 \\ b_2 \\ b_3 \end{pmatrix}$.

Trigonometric formulae:
$$\sin(A \pm B) = \sin A \cos B \pm \cos A \sin B$$
$$\cos(A \pm B) = \cos A \cos B \mp \sin A \sin B$$
$$\sin 2A = 2 \sin A \cos A$$
$$\cos 2A = \cos^2 A - \sin^2 A = 2\cos^2 A - 1 = 1 - 2\sin^2 A$$

Table of standard derivatives and integrals:

$f(x)$	$f'(x)$
$\sin ax$	$a \cos ax$
$\cos ax$	$-a \sin ax$

$f(x)$	$\int f(x)\, dx$
$\sin ax$	$-\frac{1}{a} \cos ax + C$
$\cos ax$	$\frac{1}{a} \sin ax + C$

Marks

1. On the coordinate diagram shown, A is the point (6, 8) and B is the point (12, −5). Angle AOC = p and angle COB = q.

Find the exact value of $\sin(p + q)$.

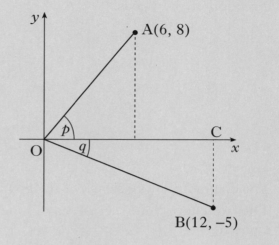

4

2. A sketch of the graph of $y = f(x)$ where $f(x) = x^3 - 6x^2 + 9x$ is shown below. The graph has a maximum at A and a minimum at B(3, 0).

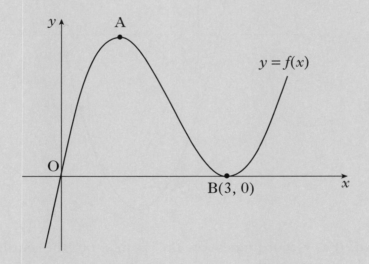

(a) Find the coordinates of the turning point at A. — 4

(b) Hence sketch the graph of $y = g(x)$ where $g(x) = f(x + 2) + 4$.
Indicate the coordinates of the turning points. There is no need to calculate the coordinates of the points of intersection with the axes. — 2

(c) Write down the range of values of k for which $g(x) = k$ has 3 real roots. — 1

[Turn over

Marks

3. Find the size of the angle $a°$ that the line joining the points A(0, −1) and B(3√3, 2) makes with the positive direction of the x-axis.

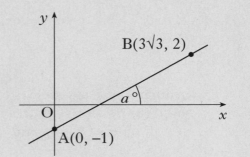

3

4. The diagram shows a sketch of the graphs of $y = 5x^2 - 15x - 8$ and $y = x^3 - 12x + 1$. The two curves intersect at A and touch at B, ie at B the curves have a common tangent.

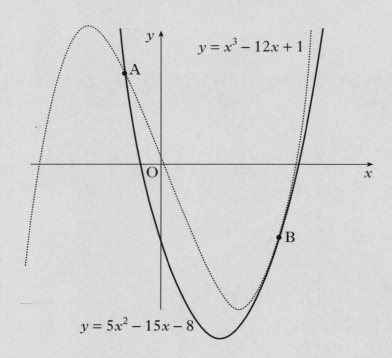

(a) (i) Find the x-coordinates of the points on the curves where the gradients are equal.

4

(ii) By considering the corresponding y-coordinates, or otherwise, distinguish geometrically between the two cases found in part (i).

1

(b) The point A is (−1, 12) and B is (3, −8).
Find the area enclosed between the two curves.

5

5. Two sequences are generated by the recurrence relations $u_{n+1} = au_n + 10$ and $v_{n+1} = a^2 v_n + 16$.

The two sequences approach the same limit as $n \to \infty$.

Determine the value of a and evaluate the limit.

6. For what range of values of k does the equation $x^2 + y^2 + 4kx - 2ky - k - 2$ represent a circle?

7. VABCD is a pyramid with a rectangular base ABCD.

Relative to some appropriate axes,

$\vec{VA}$ represents $-7\mathbf{i} - 13\mathbf{j} - 11\mathbf{k}$

$\vec{AB}$ represents $6\mathbf{i} + 6\mathbf{j} - 6\mathbf{k}$

$\vec{AD}$ represents $8\mathbf{i} - 4\mathbf{j} + 4\mathbf{k}$.

K divides BC in the ratio 1:3.

Find $\vec{VK}$ in component form.

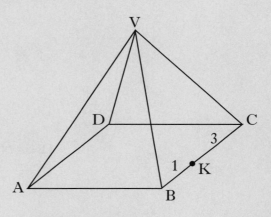

8. The graph of $y = f(x)$ passes through the point $\left(\dfrac{\pi}{9}, 1\right)$.

If $f'(x) = \sin(3x)$ express y in terms of x.

9. Evaluate $\log_5 2 + \log_5 50 - \log_5 4$.

10. Find the maximum value of $\cos x - \sin x$ and the value of x for which it occurs in the interval $0 \le x \le 2\pi$.

[END OF QUESTION PAPER]

X056/302

NATIONAL
QUALIFICATIONS
2000

THURSDAY, 25 MAY
10.30 AM – 12.00 NOON

**MATHEMATICS
HIGHER
Paper 2
Units 1, 2 and 3**

Read Carefully

1 **Calculators may be used in this paper.**

2 Full credit will be given only where the solution contains appropriate working.

3 Answers obtained by readings from scale drawings will not receive any credit.

FORMULAE LIST

Circle:

The equation $x^2 + y^2 + 2gx + 2fy + c = 0$ represents a circle centre $(-g, -f)$ and radius $\sqrt{g^2 + f^2 - c}$.

The equation $(x - a)^2 + (y - b)^2 = r^2$ represents a circle centre (a, b) and radius r.

Scalar Product: $\quad \mathbf{a}.\mathbf{b} = |\mathbf{a}| |\mathbf{b}| \cos \theta$, where θ is the angle between $\mathbf{a}$ and $\mathbf{b}$

$\quad$ or $\quad \mathbf{a}.\mathbf{b} = a_1 b_1 + a_2 b_2 + a_3 b_3$ where $\mathbf{a} = \begin{pmatrix} a_1 \\ a_2 \\ a_3 \end{pmatrix}$ and $\mathbf{b} = \begin{pmatrix} b_1 \\ b_2 \\ b_3 \end{pmatrix}$.

Trigonometric formulae:
$\sin (A \pm B) = \sin A \cos B \pm \cos A \sin B$
$\cos (A \pm B) = \cos A \cos B \mp \sin A \sin B$
$\sin 2A = 2 \sin A \cos A$
$\cos 2A = \cos^2 A - \sin^2 A = 2\cos^2 A - 1 = 1 - 2\sin^2 A$

Table of standard derivatives and integrals:

$f(x)$	$f'(x)$
$\sin ax$	$a \cos ax$
$\cos ax$	$-a \sin ax$

$f(x)$	$\int f(x)\, dx$
$\sin ax$	$-\frac{1}{a} \cos ax + C$
$\cos ax$	$\frac{1}{a} \sin ax + C$

Marks

1. The diagram shows a sketch of the graph of $y = x^3 - 3x^2 + 2x$.

    (a) Find the equation of the tangent to this curve at the point where $x = 1$.

    5

    (b) The tangent at the point (2, 0) has equation $y = 2x - 4$. Find the coordinates of the point where this tangent meets the curve again.

    5

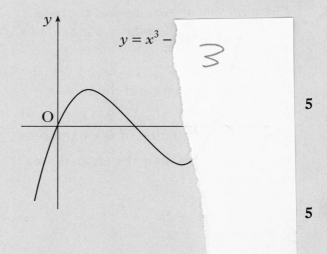

2. (a) Find the equation of AB, the perpendicular bisector of the line joining the points P(–3, 1) and Q(1, 9).

    4

    (b) C is the centre of a circle passing through P and Q. Given that QC is parallel to the y-axis, determine the equation of the circle.

    3

    (c) The tangents at P and Q intersect at T.

    Write down

    (i) the equation of the tangent at Q

    (ii) the coordinates of T.

    2

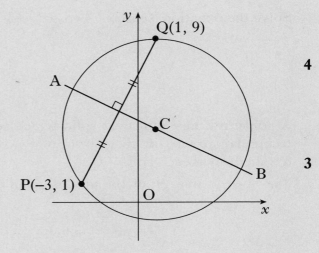

3. $f(x) = 3 - x$ and $g(x) = \frac{3}{x}$, $x \neq 0$.

    (a) Find $p(x)$ where $p(x) = f(g(x))$.

    2

    (b) If $q(x) = \frac{3}{3-x}$, $x \neq 3$, find $p(q(x))$ in its simplest form.

    3

Marks

4. The parabola shown crosses the x-axis at $(0, 0)$ and $(4, 0)$, and has a maximum at $(2, 4)$.

    The shaded area is bounded by the parabola, the x-axis and the lines $x = 2$ and $x = k$.

    (a) Find the equation of the parabola. **2**

    (b) Hence show that the shaded area, A, is given by

    $$A = -\tfrac{1}{3}k^3 + 2k^2 - \tfrac{16}{3}$$ **3**

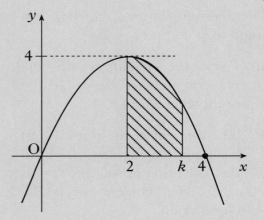

5. Solve the equation $3 \cos 2x° + \cos x° = -1$ in the interval $0 \leq x \leq 360$. **5**

6. A goldsmith has built up a solid which consists of a triangular prism of fixed volume with a regular tetrahedron at each end.

    The surface area, A, of the solid is given by

    $$A(x) = \frac{3\sqrt{3}}{2}\left(x^2 + \frac{16}{x}\right)$$

    where x is the length of each edge of the tetrahedron.

    Find the value of x which the goldsmith should use to minimise the amount of gold plating required to cover the solid. **6**

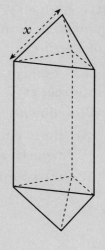

Marks

7. For what value of t are the vectors $u = \begin{pmatrix} t \\ -2 \\ 3 \end{pmatrix}$ and $v = \begin{pmatrix} 2 \\ 10 \\ t \end{pmatrix}$ perpendicular? 2

8. Given that $f(x) = (5x - 4)^{\frac{1}{2}}$, evaluate $f'(4)$. 3

9. A cuboid measuring 11 cm by 5 cm by 7 cm is placed centrally on top of another cuboid measuring 17 cm by 9 cm by 8 cm.

   Coordinate axes are taken as shown.

   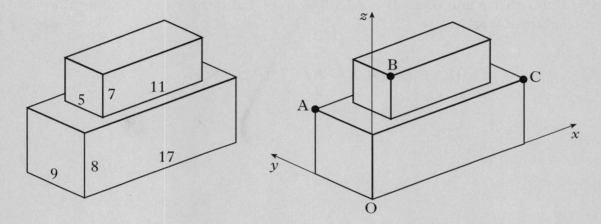

   (a) The point A has coordinates (0, 9, 8) and C has coordinates (17, 0, 8).
   Write down the coordinates of B. 1

   (b) Calculate the size of angle ABC. 6

[Turn over

10. Find $\int \dfrac{1}{(7-3x)^2}\,dx$.

Marks
2

11. The results of an experiment give rise to the graph shown.

(a) Write down the equation of the line in terms of P and Q.

2

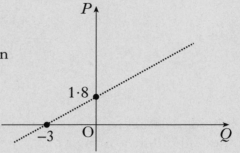

It is given that $P = \log_e p$ and $Q = \log_e q$.

(b) Show that p and q satisfy a relationship of the form $p = aq^b$, stating the values of a and b.

4

[END OF QUESTION PAPER]

2001 HIGHER

X056/301

NATIONAL
QUALIFICATIONS
2001

THURSDAY, 17 MAY
9.00 AM – 10.10 AM

MATHEMATICS
HIGHER
Units 1, 2 and 3
Paper 1
(Non-calculator)

Read Carefully

1 Calculators may **NOT** be used in this paper.

2 Full credit will be given only where the solution contains appropriate working.

3 Answers obtained by readings from scale drawings will not receive any credit.

FORMULAE LIST

Circle:

The equation $x^2 + y^2 + 2gx + 2fy + c = 0$ represents a circle centre $(-g, -f)$ and radius $\sqrt{g^2 + f^2 - c}$.

The equation $(x - a)^2 + (y - b)^2 = r^2$ represents a circle centre (a, b) and radius r.

Scalar Product: $\quad a.b = |a| |b| \cos \theta$, where θ is the angle between a and b

or $\quad a.b = a_1 b_1 + a_2 b_2 + a_3 b_3$ where $a = \begin{pmatrix} a_1 \\ a_2 \\ a_3 \end{pmatrix}$ and $b = \begin{pmatrix} b_1 \\ b_2 \\ b_3 \end{pmatrix}$.

Trigonometric formulae:

$\sin (A \pm B) = \sin A \cos B \pm \cos A \sin B$
$\cos (A \pm B) = \cos A \cos B \mp \sin A \sin B$
$\sin 2A = 2 \sin A \cos A$
$\cos 2A = \cos^2 A - \sin^2 A$
$\qquad = 2\cos^2 A - 1$
$\qquad = 1 - 2\sin^2 A$

Table of standard derivatives:

$f(x)$	$f'(x)$
$\sin ax$	$a \cos ax$
$\cos ax$	$-a \sin ax$

Table of standard integrals:

$f(x)$	$\int f(x)\, dx$
$\sin ax$	$-\frac{1}{a} \cos ax + C$
$\cos ax$	$\frac{1}{a} \sin ax + C$

ALL questions should be attempted.

Marks

1. Find the equation of the straight line which is parallel to the line with equation $2x + 3y = 5$ and which passes through the point $(2, -1)$.

2. For what value of k does the equation $x^2 - 5x + (k + 6) = 0$ have equ...

3. (a) Roadmakers look along the tops of a set of T-rods to ensure that straight sections of road are being created. Relative to suitable axes the top left corners of the T-rods are the points $A(-8, -10, -2)$, $B(-2, -1, 1)$ and $C(6, 11, 5)$.

   Determine whether or not the section of road ABC has been built in a straight line. 3

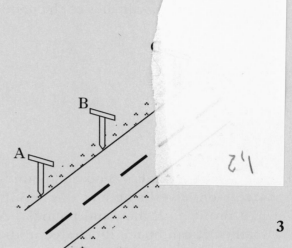

   (b) A further T-rod is placed such that D has coordinates $(1, -4, 4)$.

   Show that DB is perpendicular to AB. 3

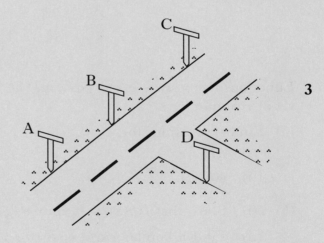

4. Given $f(x) = x^2 + 2x - 8$, express $f(x)$ in the form $(x + a)^2 - b$. 2

[Turn over

	Marks

5. (a) Solve the equation $\sin 2x° - \cos x° = 0$ in the interval $0 \leq x \leq 180$. **4**

(b) The diagram shows parts of two trigonometric graphs, $y = \sin 2x°$ and $y = \cos x°$.

Use your solutions in (a) to write down the coordinates of the point P. **1**

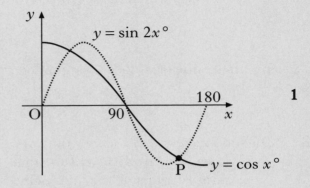

6. A company spends x thousand pounds a year on advertising and this results in a profit of P thousand pounds. A mathematical model, illustrated in the diagram, suggests that P and x are related by $P = 12x^3 - x^4$ for $0 \leq x \leq 12$.

Find the value of x which gives the maximum profit. **5**

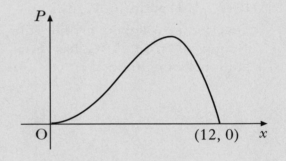

7. Functions $f(x) = \sin x$, $g(x) = \cos x$ and $h(x) = x + \frac{\pi}{4}$ are defined on a suitable set of real numbers.

(a) Find expressions for:
  (i) $f(h(x))$;
  (ii) $g(h(x))$. **2**

(b) (i) Show that $f(h(x)) = \frac{1}{\sqrt{2}}\sin x + \frac{1}{\sqrt{2}}\cos x$.

  (ii) Find a similar expression for $g(h(x))$ and hence solve the equation $f(h(x)) - g(h(x)) = 1$ for $0 \leq x \leq 2\pi$. **5**

8. Find x if $4\log_x 6 - 2\log_x 4 = 1$. **3**

9. The diagram shows the graphs of two quadratic functions $y = f(x)$ and $y = g(x)$. Both graphs have a minimum turning point at (3, 2).

Sketch the graph of $y = f'(x)$ and on the same diagram sketch the graph of $y = g'(x)$.

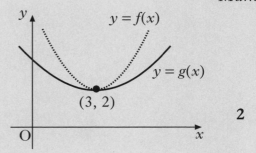

2

10. The diagram shows a sketch of part of the graph of $y = \log_2(x)$.

(a) State the values of a and b.

(b) Sketch the graph of $y = \log_2(x + 1) - 3$.

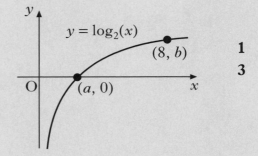

1

3

11. Circle P has equation $x^2 + y^2 - 8x - 10y + 9 = 0$. Circle Q has centre $(-2, -1)$ and radius $2\sqrt{2}$.

(a) (i) Show that the radius of circle P is $4\sqrt{2}$.

(ii) Hence show that circles P and Q touch.

4

(b) Find the equation of the tangent to circle Q at the point $(-4, 1)$.

3

(c) The tangent in (b) intersects circle P in two points. Find the x-coordinates of the points of intersection, expressing your answers in the form $a \pm b\sqrt{3}$.

3

[END OF QUESTION PAPER]

[BLANK PAGE]

X056/303

NATIONAL QUALIFICATIONS 2001

THURSDAY, 17 MAY 10.30 AM – 12.00 NOON

MATHEMATICS HIGHER
Units 1, 2 and 3
Paper 2

Read Carefully

1 **Calculators may be used in this paper.**

2 Full credit will be given only where the solution contains appropriate working.

3 Answers obtained by readings from scale drawings will not receive any credit.

FORMULAE LIST

Circle:

The equation $x^2 + y^2 + 2gx + 2fy + c = 0$ represents a circle centre $(-g, -f)$ and radius $\sqrt{g^2 + f^2 - c}$.

The equation $(x - a)^2 + (y - b)^2 = r^2$ represents a circle centre (a, b) and radius r.

Scalar Product: $\mathbf{a}.\mathbf{b} = |\mathbf{a}||\mathbf{b}| \cos \theta$, where θ is the angle between $\mathbf{a}$ and $\mathbf{b}$

or $\mathbf{a}.\mathbf{b} = a_1 b_1 + a_2 b_2 + a_3 b_3$ where $\mathbf{a} = \begin{pmatrix} a_1 \\ a_2 \\ a_3 \end{pmatrix}$ and $\mathbf{b} = \begin{pmatrix} b_1 \\ b_2 \\ b_3 \end{pmatrix}$.

Trigonometric formulae:
$$\sin(A \pm B) = \sin A \cos B \pm \cos A \sin B$$
$$\cos(A \pm B) = \cos A \cos B \mp \sin A \sin B$$
$$\sin 2A = 2 \sin A \cos A$$
$$\cos 2A = \cos^2 A - \sin^2 A$$
$$= 2\cos^2 A - 1$$
$$= 1 - 2\sin^2 A$$

Table of standard derivatives:

$f(x)$	$f'(x)$
$\sin ax$	$a \cos ax$
$\cos ax$	$-a \sin ax$

Table of standard integrals:

$f(x)$	$\int f(x)\,dx$
$\sin ax$	$-\frac{1}{a} \cos ax + C$
$\cos ax$	$\frac{1}{a} \sin ax + C$

ALL questions should be attempted.

Marks

1. (a) Given that $x + 2$ is a factor of $2x^3 + x^2 + kx + 2$, find the value of k. **3**

   (b) Hence solve the equation $2x^3 + x^2 + kx + 2 = 0$ when k takes this value. **2**

2. A curve has equation $y = x - \dfrac{16}{\sqrt{x}}$, $x > 0$.

   Find the equation of the tangent at the point where $x = 4$. **6**

3. On the first day of March, a bank loans a man £2500 at a fixed rate of interest of 1·5% per month. This interest is added on the last day of each month and is calculated on the amount due on the first day of the month. He agrees to make repayments on the first day of each subsequent month. Each repayment is £300 except for the smaller final amount which will pay off the loan.

   (a) The amount that he owes at the start of each month is taken to be the amount still owing just after the monthly repayment has been made.

   Let u_n and u_{n+1} represent the amounts that he owes at the starts of two successive months. Write down a recurrence relation involving u_{n+1} and u_n. **2**

   (b) Find the date and the amount of the final payment. **4**

4. A box in the shape of a cuboid is designed with **circles** of different sizes on each face.

   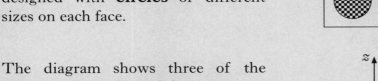

   The diagram shows three of the circles, where the origin represents one of the corners of the cuboid. The centres of the circles are A(6, 0, 7), B(0, 5, 6) and C(4, 5, 0).

   Find the size of angle ABC. **7**

   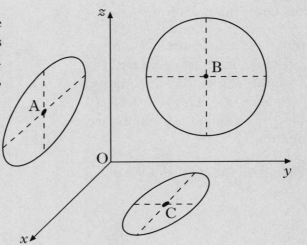

   [Turn over

Marks

5. Express $8\cos x° - 6\sin x°$ in the form $k\cos(x+a)°$ where $k > 0$ and $0 < a < 360$. 4

6. Find $\displaystyle\int \frac{(x^2 - 2)(x^2 + 2)}{x^2} dx$, $x \neq 0$ 4

7. Triangle ABC has vertices A(2, 2), B(12, 2) and C(8, 6).

    (a) Write down the equation of l_1, the perpendicular bisector of AB. 1

    (b) Find the equation of l_2, the perpendicular bisector of AC. 4

    (c) Find the point of intersection of lines l_1 and l_2. 1

    (d) Hence find the equation of the circle passing through A, B and C. 2

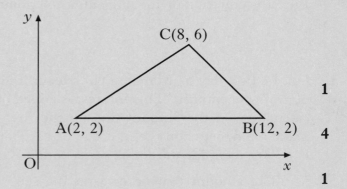

8. A firm asked for a logo to be designed involving the letters A and U. Their initial sketch is shown in the hexagon.

    A mathematical representation of the final logo is shown in the coordinate diagram.

    The curve has equation $y = (x + 1)(x - 1)(x - 3)$ and the straight line has equation $y = 5x - 5$. The point (1, 0) is the centre of half-turn symmetry.

    Calculate the total shaded area. 7

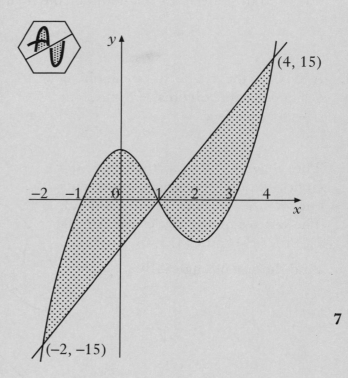

Marks

9. Before a forest fire was brought under control, the spread of the fire was described by a law of the form $A = A_0 e^{kt}$ where A_0 is the area covered by the fire when it was first detected and A is the area covered by the fire t hours later.

   If it takes one and half hours for the area of the forest fire to double, find the value of the constant k. **3**

10. A curve for which $\dfrac{dy}{dx} = 3\sin(2x)$ passes through the point $\left(\tfrac{5}{12}\pi, \sqrt{3}\right)$.

    Find y in terms of x. **4**

11. The diagram shows a sketch of a parabola passing through $(-1, 0)$, $(0, p)$ and $(p, 0)$.

    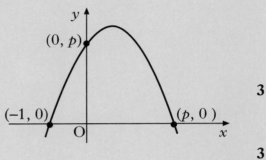

    (a) Show that the equation of the parabola is $y = p + (p-1)x - x^2$. **3**

    (b) For what value of p will the line $y = x + p$ be a tangent to this curve? **3**

[END OF QUESTION PAPER]

[BLANK PAGE]

2002 HIGHER

X100/301

NATIONAL QUALIFICATIONS 2002

MONDAY, 27 MAY 9.00 AM – 10.10 AM

MATHEMATICS HIGHER
Units 1, 2 and 3
Paper 1
(Non-calculator)

Read Carefully

1. **Calculators may NOT be used in this paper.**
2. Full credit will be given only where the solution contains appropriate working.
3. Answers obtained by readings from scale drawings will not receive any credit.

FORMULAE LIST

Circle:

The equation $x^2 + y^2 + 2gx + 2fy + c = 0$ represents a circle centre $(-g, -f)$ and radius $\sqrt{g^2 + f^2 - c}$.

The equation $(x - a)^2 + (y - b)^2 = r^2$ represents a circle centre (a, b) and radius r.

Scalar Product: $\quad \mathbf{a}.\mathbf{b} = |\mathbf{a}| |\mathbf{b}| \cos \theta$, where θ is the angle between $\mathbf{a}$ and $\mathbf{b}$

or $\quad \mathbf{a}.\mathbf{b} = a_1 b_1 + a_2 b_2 + a_3 b_3$ where $\mathbf{a} = \begin{pmatrix} a_1 \\ a_2 \\ a_3 \end{pmatrix}$ and $\mathbf{b} = \begin{pmatrix} b_1 \\ b_2 \\ b_3 \end{pmatrix}$.

Trigonometric formulae:

$$\sin(A \pm B) = \sin A \cos B \pm \cos A \sin B$$
$$\cos(A \pm B) = \cos A \cos B \mp \sin A \sin B$$
$$\sin 2A = 2 \sin A \cos A$$
$$\cos 2A = \cos^2 A - \sin^2 A$$
$$= 2\cos^2 A - 1$$
$$= 1 - 2\sin^2 A$$

Table of standard derivatives:

$f(x)$	$f'(x)$
$\sin ax$	$a \cos ax$
$\cos ax$	$-a \sin ax$

Table of standard integrals:

$f(x)$	$\int f(x)\, dx$
$\sin ax$	$-\frac{1}{a} \cos ax + C$
$\cos ax$	$\frac{1}{a} \sin ax + C$

ALL questions should be attempted.

Marks

1. The point P(2, 3) lies on the circle $(x + 1)^2 + (y - 1)^2 = 13$. Find the equation of the tangent at P.   4

2. The point Q divides the line joining P(−1, −1, 0) to R(5, 2, −3) in the ratio 2 : 1. Find the coordinates of Q.   3

3. Functions f and g are defined on suitable domains by $f(x) = \sin(x°)$ and $g(x) = 2x$.

   (a) Find expressions for:
      (i) $f(g(x))$;
      (ii) $g(f(x))$.   2

   (b) Solve $2f(g(x)) = g(f(x))$ for $0 \le x \le 360$.   5

4. Find the coordinates of the point on the curve $y = 2x^2 - 7x + 10$ where the tangent to the curve makes an angle of 45° with the positive direction of the x-axis.   4

5. In triangle ABC, show that the exact value of $\sin(a + b)$ is $\dfrac{2}{\sqrt{5}}$.   4

6. The graph of a function f intersects the x-axis at $(-a, 0)$ and $(e, 0)$ as shown.

There is a point of inflexion at $(0, b)$ and a maximum turning point at (c, d).

Sketch the graph of the derived function f'.   3

[Turn over for Questions 7 to 11 on *Page four*

		Marks
7.	(a) Express $f(x) = x^2 - 4x + 5$ in the form $f(x) = (x-a)^2 + b$.	2
	(b) On the same diagram sketch:	
	(i) the graph of $y = f(x)$;	
	(ii) the graph of $y = 10 - f(x)$.	4
	(c) Find the range of values of x for which $10 - f(x)$ is positive.	1

8. The diagram shows the graph of a cosine function from 0 to π.

   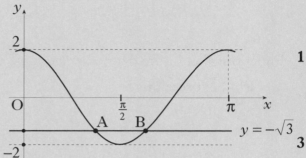

   (a) State the equation of the graph.   1

   (b) The line with equation $y = -\sqrt{3}$ intersects this graph at points A and B.
   Find the coordinates of B.   3

9. (a) Write $\sin(x) - \cos(x)$ in the form $k\sin(x - a)$ stating the values of k and a where $k > 0$ and $0 \leq a \leq 2\pi$.   4

   (b) Sketch the graph of $y = \sin(x) - \cos(x)$ for $0 \leq x \leq 2\pi$, showing clearly the graph's maximum and minimum values and where it cuts the x-axis and the y-axis.   3

10. (a) Find the derivative of the function $f(x) = (8 - x^3)^{\frac{1}{2}}$, $x < 2$.   2

    (b) Hence write down $\displaystyle\int \frac{x^2}{(8-x^3)^{\frac{1}{2}}} dx$.   1

11. The graph illustrates the law $y = kx^n$.
    If the straight line passes through A(0·5, 0) and B(0, 1), find the values of k and n.   4

    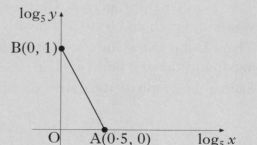

[END OF QUESTION PAPER]

X100/303

NATIONAL QUALIFICATIONS 2002

MONDAY, 27 MAY 10.30 AM – 12.00 NOON

MATHEMATICS HIGHER
Units 1, 2 and 3
Paper 2

Read Carefully

1 **Calculators may be used in this paper.**
2 Full credit will be given only where the solution contains appropriate working.
3 Answers obtained by readings from scale drawings will not receive any credit.

FORMULAE LIST

Circle:

The equation $x^2 + y^2 + 2gx + 2fy + c = 0$ represents a circle centre $(-g, -f)$ and radius $\sqrt{g^2 + f^2 - c}$.

The equation $(x - a)^2 + (y - b)^2 = r^2$ represents a circle centre (a, b) and radius r.

Scalar Product: $\quad a.b = |a| |b| \cos \theta$, where θ is the angle between a and b

or $\quad a.b = a_1 b_1 + a_2 b_2 + a_3 b_3$ where $a = \begin{pmatrix} a_1 \\ a_2 \\ a_3 \end{pmatrix}$ and $b = \begin{pmatrix} b_1 \\ b_2 \\ b_3 \end{pmatrix}$.

Trigonometric formulae:

$$\sin(A \pm B) = \sin A \cos B \pm \cos A \sin B$$
$$\cos(A \pm B) = \cos A \cos B \mp \sin A \sin B$$
$$\sin 2A = 2 \sin A \cos A$$
$$\cos 2A = \cos^2 A - \sin^2 A$$
$$= 2\cos^2 A - 1$$
$$= 1 - 2\sin^2 A$$

Table of standard derivatives:

$f(x)$	$f'(x)$
$\sin ax$	$a \cos ax$
$\cos ax$	$-a \sin ax$

Table of standard integrals:

$f(x)$	$\int f(x)\, dx$
$\sin ax$	$-\frac{1}{a} \cos ax + C$
$\cos ax$	$\frac{1}{a} \sin ax + C$

ALL questions should be attempted.

Marks

1. Triangle ABC has vertices A(–1, 6), B(–3, –2) and C(5, 2).

   Find

   (a) the equation of the line p, the median from C of triangle ABC. **3**

   (b) the equation of the line q, the perpendicular bisector of BC. **4**

   (c) the coordinates of the point of intersection of the lines p and q. **1**

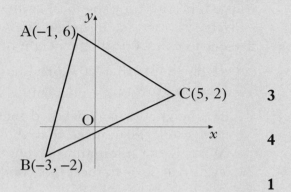

2. The diagram shows a square-based pyramid of height 8 units.

   Square OABC has a side length of 6 units.

   The coordinates of A and D are (6, 0, 0) and (3, 3, 8).

   C lies on the y-axis.

   (a) Write down the coordinates of B. **1**

   (b) Determine the components of $\vec{DA}$ and $\vec{DB}$. **2**

   (c) Calculate the size of angle ADB. **4**

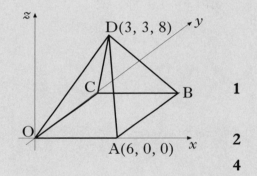

3. The diagram shows part of the graph of the curve with equation $y = 2x^3 - 7x^2 + 4x + 4$.

   (a) Find the x-coordinate of the maximum turning point. **5**

   (b) Factorise $2x^3 - 7x^2 + 4x + 4$. **3**

   (c) State the coordinates of the point A and hence find the values of x for which $2x^3 - 7x^2 + 4x + 4 < 0$. **2**

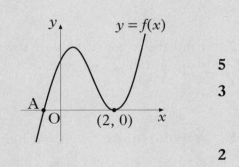

[Turn over

Marks

4. A man decides to plant a number of fast-growing trees as a boundary between his property and the property of his next door neighbour. He has been warned, however, by the local garden centre that, during any year, the trees are expected to increase in height by 0·5 metres. In response to this warning he decides to trim 20% off the height of the trees at the start of any year.

   (a) If he adopts the "20% pruning policy", to what height will he expect the trees to grow in the long run? 3

   (b) His neighbour is concerned that the trees are growing at an alarming rate and wants assurances that the trees will grow no taller than 2 metres. What is the minimum percentage that the trees will need to be trimmed each year so as to meet this condition? 3

5. Calculate the shaded area enclosed between the parabolas with equations $y = 1 + 10x - 2x^2$ and $y = 1 + 5x - x^2$. 6

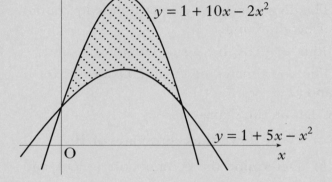

6. Find the equation of the tangent to the curve $y = 2\sin\left(x - \dfrac{\pi}{6}\right)$ at the point where $x = \dfrac{\pi}{3}$. 4

7. Find the x-coordinate of the point where the graph of the curve with equation $y = \log_3(x - 2) + 1$ intersects the x-axis. 3

8. A point moves in a straight line such that its acceleration a is given by $a = 2(4 - t)^{\frac{1}{2}}$, $0 \le t \le 4$. If it starts at rest, find an expression for the velocity v where $a = \dfrac{dv}{dt}$. 4

9. Show that the equation $(1 - 2k)x^2 - 5kx - 2k = 0$ has real roots for all integer values of k. 5

10. The shaded rectangle on this map represents the planned extension to the village hall. It is hoped to provide the largest possible area for the extension.

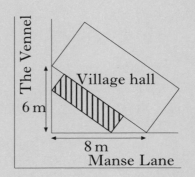

The coordinate diagram represents the right angled triangle of ground behind the hall. The extension has length l metres and breadth b metres, as shown. One corner of the extension is at the point $(a, 0)$.

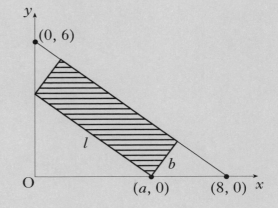

(a) (i) Show that $l = \tfrac{5}{4}a$.

(ii) Express b in terms of a and hence deduce that the area, $A\,\text{m}^2$, of the extension is given by $A = \tfrac{3}{4}a(8-a)$.   **3**

(b) Find the value of a which produces the largest area of the extension.   **4**

[END OF QUESTION PAPER]

2002 HIGHER WINTER DIET

Official SQA Past Papers: Higher Mathematics

W100/301

NATIONAL QUALIFICATIONS 2002

FRIDAY, 18 JANUARY
9.00 AM – 10.10 AM

MATHEMATICS HIGHER
Units 1, 2 and 3
Paper 1
(Non-calculator)

Read Carefully

1 Calculators may **NOT** be used in this paper.
2 Full credit will be given only where the solution contains appropriate working.
3 Answers obtained by readings from scale drawings will not receive any credit.

FORMULAE LIST

Circle:

The equation $x^2 + y^2 + 2gx + 2fy + c = 0$ represents a circle centre $(-g, -f)$ and radius $\sqrt{g^2 + f^2 - c}$.

The equation $(x - a)^2 + (y - b)^2 = r^2$ represents a circle centre (a, b) and radius r.

Scalar Product: $\quad \mathbf{a.b} = |\mathbf{a}| |\mathbf{b}| \cos \theta$, where θ is the angle between $\mathbf{a}$ and $\mathbf{b}$

or $\quad \mathbf{a.b} = a_1 b_1 + a_2 b_2 + a_3 b_3$ where $\mathbf{a} = \begin{pmatrix} a_1 \\ a_2 \\ a_3 \end{pmatrix}$ and $\mathbf{b} = \begin{pmatrix} b_1 \\ b_2 \\ b_3 \end{pmatrix}$.

Trigonometric formulae:

$\sin(A \pm B) = \sin A \cos B \pm \cos A \sin B$
$\cos(A \pm B) = \cos A \cos B \mp \sin A \sin B$
$\sin 2A = 2 \sin A \cos A$
$\cos 2A = \cos^2 A - \sin^2 A$
$ = 2\cos^2 A - 1$
$ = 1 - 2\sin^2 A$

Table of standard derivatives:

$f(x)$	$f'(x)$
$\sin ax$	$a \cos ax$
$\cos ax$	$-a \sin ax$

Table of standard integrals:

$f(x)$	$\int f(x)\,dx$
$\sin ax$	$-\frac{1}{a} \cos ax + C$
$\cos ax$	$\frac{1}{a} \sin ax + C$

ALL questions should be attempted.

Marks

1. (a) Find the equation of the straight line through the points A(−1, 5) and B(3, 1). **2**

   (b) Find the size of the angle which AB makes with the positive direction of the *x*-axis. **2**

2. (a) If $u = \begin{pmatrix} 1 \\ 7 \\ -2 \end{pmatrix}$ and $v = \begin{pmatrix} 1 \\ -2 \\ 1 \end{pmatrix}$, write down the components of $u + 3v$ and $u - 3v$. **2**

   (b) Hence, or otherwise, show that $u + 3v$ and $u - 3v$ are perpendicular. **2**

3. Find the equation of the tangent to the curve with equation $y = \frac{3}{x}$ at the point P where $x = 1$. **5**

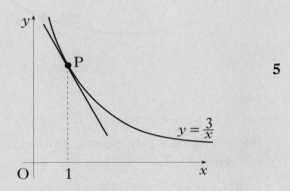

4. (a) Write down the exact values of $\sin\left(\frac{\pi}{3}\right)$ and $\cos\left(\frac{\pi}{3}\right)$. **1**

   (b) If $\tan x = 4 \sin\left(\frac{\pi}{3}\right) \cos\left(\frac{\pi}{3}\right)$, find the exact values of x for $0 \leq x \leq 2\pi$. **2**

5. Given that $(x − 2)$ and $(x + 3)$ are factors of $f(x)$ where $f(x) = 3x^3 + 2x^2 + cx + d$, find the values of c and d. **5**

[Turn over

Marks

6. The side view of part of a roller coaster ride is shown by the path PQRS.
The curve PQ is an arc of the circle with equation $x^2 + y^2 + 4x - 10y + 9 = 0$.
The curve QRS is part of the parabola with equation $y = -x^2 + 6x - 5$.
The point Q has coordinates (2, 3).

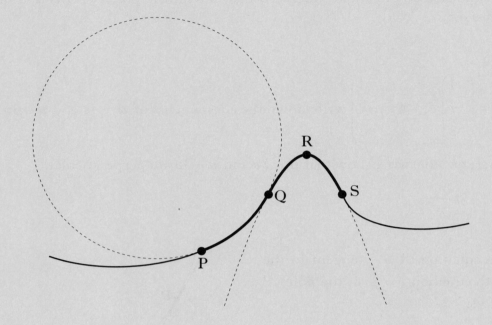

(a) Find the equation of the tangent to the circle at Q.   4

(b) Show that this tangent to the circle at Q is also the tangent to the parabola at Q.   2

7. Find $\int \left(\sqrt[3]{x} - \dfrac{1}{\sqrt{x}} \right) dx$.   4

8. The diagram shows part of the graph of $y = 2^x$.

(a) Sketch the graph of $y = 2^{-x} - 8$.   2

(b) Find the coordinates of the points where it crosses the x and y axes.   2

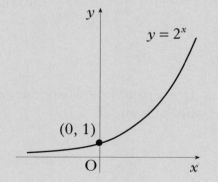

  Marks

9. The function f, defined on a suitable domain, is given by $f(x) = \frac{3}{x+1}$.

   (a) Find an expression for $h(x)$ where $h(x) = f(f(x))$, giving your answer as a fraction in its simplest form.   3

   (b) Describe any restriction on the domain of h.   1

10. A function f is defined by $f(x) = 2x + 3 + \frac{18}{x-4}$, $x \neq 4$.

    Find the values of x for which the function is increasing.   5

11. PQRSTU is a regular hexagon of side 2 units. $\vec{PQ}$, $\vec{QR}$ and $\vec{RS}$ represent vectors $\mathbf{a}$, $\mathbf{b}$ and $\mathbf{c}$ respectively.

    Find the value of $\mathbf{a}.(\mathbf{b} + \mathbf{c})$.   3

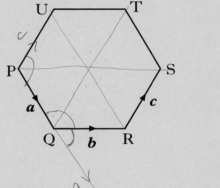

12. If $\log_a p = \cos^2 x$ and $\log_a r = \sin^2 x$, show that $pr = a$.   3

[END OF QUESTION PAPER]

W100/303

NATIONAL QUALIFICATIONS 2002

FRIDAY, 18 JANUARY 10.30 AM – 12.00 NOON

MATHEMATICS
HIGHER
Units 1, 2 and 3
Paper 2

Read Carefully

1 **Calculators may be used in this paper.**
2 Full credit will be given only where the solution contains appropriate working.
3 Answers obtained by readings from scale drawings will not receive any credit.

FORMULAE LIST

Circle:

The equation $x^2 + y^2 + 2gx + 2fy + c = 0$ represents a circle centre $(-g, -f)$ and radius $\sqrt{g^2 + f^2 - c}$.

The equation $(x - a)^2 + (y - b)^2 = r^2$ represents a circle centre (a, b) and radius r.

Scalar Product: $\mathbf{a}.\mathbf{b} = |\mathbf{a}| |\mathbf{b}| \cos \theta$, where θ is the angle between $\mathbf{a}$ and $\mathbf{b}$

or $\mathbf{a}.\mathbf{b} = a_1 b_1 + a_2 b_2 + a_3 b_3$ where $\mathbf{a} = \begin{pmatrix} a_1 \\ a_2 \\ a_3 \end{pmatrix}$ and $\mathbf{b} = \begin{pmatrix} b_1 \\ b_2 \\ b_3 \end{pmatrix}$.

Trigonometric formulae:

$$\sin(A \pm B) = \sin A \cos B \pm \cos A \sin B$$
$$\cos(A \pm B) = \cos A \cos B \mp \sin A \sin B$$
$$\sin 2A = 2 \sin A \cos A$$
$$\cos 2A = \cos^2 A - \sin^2 A$$
$$= 2\cos^2 A - 1$$
$$= 1 - 2\sin^2 A$$

Table of standard derivatives:

$f(x)$	$f'(x)$
$\sin ax$	$a \cos ax$
$\cos ax$	$-a \sin ax$

Table of standard integrals:

$f(x)$	$\int f(x)\, dx$
$\sin ax$	$-\frac{1}{a} \cos ax + C$
$\cos ax$	$\frac{1}{a} \sin ax + C$

ALL questions should be attempted.

Marks

1. The diagram shows a rhombus PQRS with its diagonals PR and QS.

   PR has equation $y = 2x - 2$.

   Q has coordinates $(-2, 4)$.

   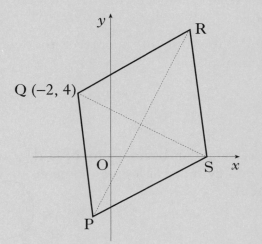

   (a) (i) Find the equation of the diagonal QS.

   (ii) Find the coordinates of T, the point of intersection of PR and QS. **6**

   (b) R is the point $(5, 8)$. Write down the coordinates of P. **2**

2. With reference to a suitable set of coordinate axes, A, B and C are the points $(-8, 10, 20)$, $(-2, 1, 8)$ and $(0, -2, 4)$ respectively.

   Show that A, B and C are collinear and find the ratio AB : BC. **4**

3. (a) Calculate the limit as $n \to \infty$ of the sequence defined by $u_{n+1} = 0 \cdot 9 u_n + 10$, $u_0 = 1$. **3**

   (b) Determine the least value of n for which u_n is greater than half of this limit and the corresponding value of u_n. **2**

4. (a) Write $\sqrt{3} \sin x° + \cos x°$ in the form $k \sin(x + a)°$ where $k > 0$ and $0 \le a < 360$. **4**

   (b) Hence find the maximum value of $5 + \sqrt{3} \sin x° + \cos x°$ and determine the corresponding value of x in the interval $0 \le x \le 360$. **2**

5. Solve the equation $\cos 2x - 2\sin^2 x = 0$ in the interval $0 \le x < 2\pi$. **4**

[Turn over

Marks

6. The graph of $f(x) = 2x^3 - 5x^2 - 3x + 1$ has been sketched in the diagram shown.

Find the value of a correct to one decimal place.

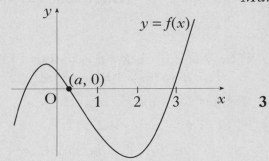

3

7. A rectangular beam is to be cut from a cylindrical log of diameter 20 cm.

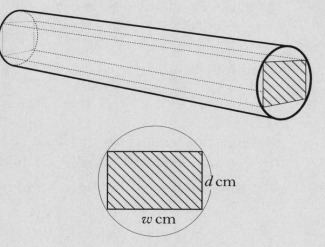

The diagram shows a cross-section of the log and beam where the beam has a breadth of w cm and a depth of d cm.

The strength S of the beam is given by

$S = 1 \cdot 7\, w\, (400 - w^2)$.

Find the dimensions of the beam for maximum strength.

5

8. Find $\int_0^1 \left(\cos(3x) - \sin\left(\tfrac{1}{3}x + 1\right) \right) dx$ correct to 3 decimal places.

3

9. A researcher modelled the size N of a colony of bacteria t hours after the beginning of her observations by $N(t) = 950 \times (2 \cdot 6)^{0 \cdot 2t}$.

(a) What was the size of the colony when observations began?

1

(b) How long does it take for the size of the colony to be multiplied by 10?

4

10. The line $y + 2x = k$, $k > 0$, is a tangent to the circle $x^2 + y^2 - 2x - 4 = 0$.

(a) Find the value of k.

7

(b) Deduce the coordinates of the point of contact.

2

Marks

11. An energy efficient building is designed with solar cells covering the whole of its south facing roof. The energy generated by the solar cells is directly proportional to the area, in square units, of the solar roof.

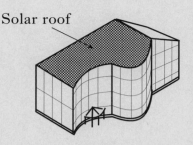

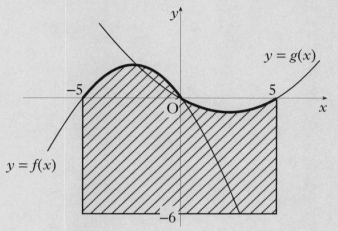

The shape of the solar roof can be represented on the coordinate plane as the shaded area bounded by the functions $f(x) = \frac{1}{4}(-x^2 - 5x)$, $g(x) = \frac{1}{12}(x^2 - 5x)$ and the lines $x = -5$, $x = 5$ and $y = -6$.

(a) Find the area of the solar roof. **7**

(b) Ten square units of solar cells generate a maximum of 1 kilowatt.

What is the maximum energy the solar roof can generate in kilowatts (to the nearest kilowatt)? **1**

[*END OF QUESTION PAPER*]

2003 HIGHER

X100/301

NATIONAL QUALIFICATIONS 2003

WEDNESDAY, 21 MAY 9.00 AM – 10.10 AM

MATHEMATICS HIGHER
Units 1, 2 and 3
Paper 1
(Non-calculator)

Read Carefully

1. **Calculators may NOT be used in this paper.**
2. Full credit will be given only where the solution contains appropriate working.
3. Answers obtained by readings from scale drawings will not receive any credit.

FORMULAE LIST

Circle:

The equation $x^2 + y^2 + 2gx + 2fy + c = 0$ represents a circle centre $(-g, -f)$ and radius $\sqrt{g^2 + f^2 - c}$.

The equation $(x - a)^2 + (y - b)^2 = r^2$ represents a circle centre (a, b) and radius r.

Scalar Product: $\quad \boldsymbol{a}.\boldsymbol{b} = |\boldsymbol{a}| |\boldsymbol{b}| \cos \theta$, where θ is the angle between $\boldsymbol{a}$ and $\boldsymbol{b}$

or $\quad \boldsymbol{a}.\boldsymbol{b} = a_1 b_1 + a_2 b_2 + a_3 b_3$ where $\boldsymbol{a} = \begin{pmatrix} a_1 \\ a_2 \\ a_3 \end{pmatrix}$ and $\boldsymbol{b} = \begin{pmatrix} b_1 \\ b_2 \\ b_3 \end{pmatrix}$.

Trigonometric formulae:

$$\sin(A \pm B) = \sin A \cos B \pm \cos A \sin B$$
$$\cos(A \pm B) = \cos A \cos B \mp \sin A \sin B$$
$$\sin 2A = 2 \sin A \cos A$$
$$\cos 2A = \cos^2 A - \sin^2 A$$
$$= 2\cos^2 A - 1$$
$$= 1 - 2\sin^2 A$$

Table of standard derivatives:

$f(x)$	$f'(x)$
$\sin ax$	$a \cos ax$
$\cos ax$	$-a \sin ax$

Table of standard integrals:

$f(x)$	$\int f(x)\, dx$
$\sin ax$	$-\frac{1}{a} \cos ax + C$
$\cos ax$	$\frac{1}{a} \sin ax + C$

ALL questions should be attempted.

1. Find the equation of the line which passes through the point $(-1, 3)$ and is perpendicular to the line with equation $4x + y - 1 = 0$. **3**

2. (a) Write $f(x) = x^2 + 6x + 11$ in the form $(x + a)^2 + b$. **2**

   (b) Hence or otherwise sketch the graph of $y = f(x)$. **2**

3. Vectors $\mathbf{u}$ and $\mathbf{v}$ are defined by $\mathbf{u} = 3\mathbf{i} + 2\mathbf{j}$ and $\mathbf{v} = 2\mathbf{i} - 3\mathbf{j} + 4\mathbf{k}$.

   Determine whether or not $\mathbf{u}$ and $\mathbf{v}$ are perpendicular to each other. **2**

4. A recurrence relation is defined by $u_{n+1} = pu_n + q$, where $-1 < p < 1$ and $u_0 = 12$.

   (a) If $u_1 = 15$ and $u_2 = 16$, find the values of p and q. **2**

   (b) Find the limit of this recurrence relation as $n \to \infty$. **2**

5. Given that $f(x) = \sqrt{x} + \dfrac{2}{x^2}$, find $f'(4)$. **5**

6. A and B are the points $(-1, -3, 2)$ and $(2, -1, 1)$ respectively.

   B and C are the points of trisection of AD, that is AB = BC = CD.

   Find the coordinates of D. **3**

   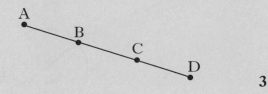

7. Show that the line with equation $y = 2x + 1$ does not intersect the parabola with equation $y = x^2 + 3x + 4$. **5**

8. Find $\displaystyle\int_0^1 \dfrac{dx}{(3x+1)^{\frac{1}{2}}}$. **4**

9. Functions $f(x) = \dfrac{1}{x-4}$ and $g(x) = 2x + 3$ are defined on suitable domains.

   (a) Find an expression for $h(x)$ where $h(x) = f(g(x))$. **2**

   (b) Write down any restriction on the domain of h. **1**

[*Turn over for Questions 10 to 12 on Page four*

Marks

10. A is the point (8, 4). The line OA is inclined at an angle p radians to the x-axis.

   (a) Find the exact values of:
   (i) $\sin(2p)$;
   (ii) $\cos(2p)$.

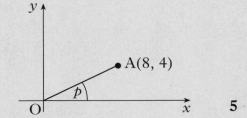

5

The line OB is inclined at an angle $2p$ radians to the x-axis.

(b) Write down the exact value of the gradient of OB.

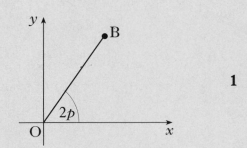

1

11.
   - O, A and B are the centres of the three circles shown in the diagram below.
   - The two outer circles are congruent and each touches the smallest circle.
   - Circle centre A has equation $(x - 12)^2 + (y + 5)^2 = 25$.
   - The three centres lie on a parabola whose axis of symmetry is shown by the broken line through A.

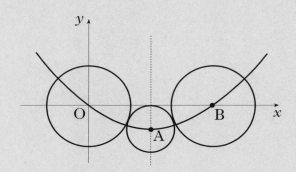

   (a) (i) State the coordinates of A and find the length of the line OA. 2
   (ii) Hence find the equation of the circle with centre B. 3

   (b) The equation of the parabola can be written in the form $y = px(x + q)$.
   Find the values of p and q. 2

12. Simplify $3 \log_e(2e) - 2 \log_e(3e)$ expressing your answer in the form $A + \log_e B - \log_e C$ where A, B and C are whole numbers. 4

[END OF QUESTION PAPER]

X100/303

NATIONAL QUALIFICATIONS 2003

WEDNESDAY, 21 MAY 10.30 AM – 12.00 NOON

MATHEMATICS
HIGHER
Units 1, 2 and 3
Paper 2

Read Carefully

1 **Calculators may be used in this paper.**
2 Full credit will be given only where the solution contains appropriate working.
3 Answers obtained by readings from scale drawings will not receive any credit.

FORMULAE LIST

Circle:

The equation $x^2 + y^2 + 2gx + 2fy + c = 0$ represents a circle centre $(-g, -f)$ and radius $\sqrt{g^2 + f^2 - c}$.

The equation $(x - a)^2 + (y - b)^2 = r^2$ represents a circle centre (a, b) and radius r.

Scalar Product: $\quad \mathbf{a}.\mathbf{b} = |\mathbf{a}||\mathbf{b}| \cos \theta$, where θ is the angle between $\mathbf{a}$ and $\mathbf{b}$

or $\quad \mathbf{a}.\mathbf{b} = a_1 b_1 + a_2 b_2 + a_3 b_3$ where $\mathbf{a} = \begin{pmatrix} a_1 \\ a_2 \\ a_3 \end{pmatrix}$ and $\mathbf{b} = \begin{pmatrix} b_1 \\ b_2 \\ b_3 \end{pmatrix}$.

Trigonometric formulae:

$\sin(A \pm B) = \sin A \cos B \pm \cos A \sin B$
$\cos(A \pm B) = \cos A \cos B \mp \sin A \sin B$
$\sin 2A = 2 \sin A \cos A$
$\cos 2A = \cos^2 A - \sin^2 A$
$\quad\quad\quad = 2\cos^2 A - 1$
$\quad\quad\quad = 1 - 2\sin^2 A$

Table of standard derivatives:

$f(x)$	$f'(x)$
$\sin ax$	$a \cos ax$
$\cos ax$	$-a \sin ax$

Table of standard integrals:

$f(x)$	$\int f(x)\, dx$
$\sin ax$	$-\frac{1}{a} \cos ax + C$
$\cos ax$	$\frac{1}{a} \sin ax + C$

ALL questions should be attempted.

Marks

1. $f(x) = 6x^3 - 5x^2 - 17x + 6$.

   (a) Show that $(x - 2)$ is a factor of $f(x)$.

   (b) Express $f(x)$ in its fully factorised form.   **4**

2. The diagram shows a sketch of part of the graph of a trigonometric function whose equation is of the form $y = a\sin(bx) + c$.

   Determine the values of a, b and c.   **3**

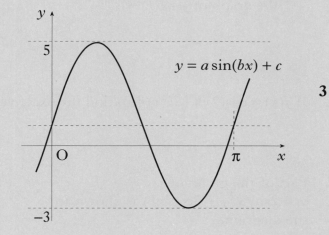

3. The incomplete graphs of $f(x) = x^2 + 2x$ and $g(x) = x^3 - x^2 - 6x$ are shown in the diagram. The graphs intersect at A(4, 24) and the origin.

   Find the shaded area enclosed between the curves.   **5**

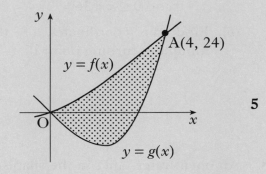

4. (a) Find the equation of the tangent to the curve with equation $y = x^3 + 2x^2 - 3x + 2$ at the point where $x = 1$.   **5**

   (b) Show that this line is also a tangent to the circle with equation $x^2 + y^2 - 12x - 10y + 44 = 0$ and state the coordinates of the point of contact.   **6**

[Turn over

5. The diagram shows the graph of a function f.

   f has a minimum turning point at $(0, -3)$ and a point of inflexion at $(-4, 2)$.

   (a) Sketch the graph of $y = f(-x)$.

   (b) On the same diagram, sketch the graph of $y = 2f(-x)$.

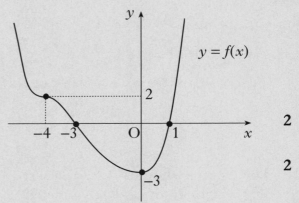

6. If $f(x) = \cos(2x) - 3\sin(4x)$, find the exact value of $f'\left(\dfrac{\pi}{6}\right)$.

7. Part of the graph of $y = 2\sin(x°) + 5\cos(x°)$ is shown in the diagram.

   (a) Express $y = 2\sin(x°) + 5\cos(x°)$ in the form $k\sin(x° + a°)$ where $k > 0$ and $0 \le a < 360$.

   (b) Find the coordinates of the minimum turning point P.

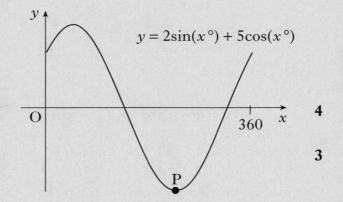

8. An open water tank, in the shape of a triangular prism, has a capacity of 108 litres. The tank is to be lined on the inside in order to make it watertight.

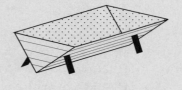

   The triangular cross-section of the tank is right-angled and isosceles, with equal sides of length x cm. The tank has a length of l cm.

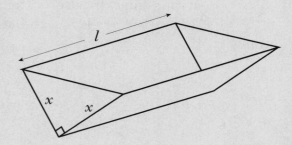

   (a) Show that the surface area to be lined, A cm^2, is given by $A(x) = x^2 + \dfrac{432000}{x}$.

   (b) Find the value of x which minimises this surface area.

Marks

9. The diagram shows vectors **a** and **b**.
   If $|a| = 5$, $|b| = 4$ and $a \cdot (a + b) = 36$, find the size of the acute angle θ between **a** and **b**. 4

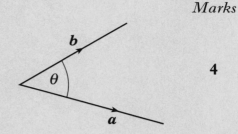

10. Solve the equation $3\cos(2x) + 10\cos(x) - 1 = 0$ for $0 \le x \le \pi$, correct to 2 decimal places. 5

11. (a) (i) Sketch the graph of $y = a^x + 1$, $a > 2$.
    (ii) On the same diagram, sketch the graph of $y = a^{x+1}$, $a > 2$. 2

    (b) Prove that the graphs intersect at a point where the x-coordinate is $\log_a\left(\dfrac{1}{a-1}\right)$. 3

[*END OF QUESTION PAPER*]

[BLANK PAGE]

[BLANK PAGE]

[BLANK PAGE]